The Art of the Amplifier

Michael Doyle

A TOME OF TONE PRODUCTION

ISBN 0-7935-3989-7

7777 W. BLUEMOUND RD. P.O. BOX 13819 MILWAUKEE, WI 53213

Copyright © 1996 by HAL LEONARD CORPORATION
International Copyright Secured   All Rights Reserved

No part of this publication may be reproduced in any form or by any
means without the prior written permission of the Publisher.

*To Cabot S. Bull and Sidney Rodda,
who, while working for EMI/MOV in 1933,
invented the "Beam (or 'Kinkless')
Tetrode" output valve*

ACKNOWLEDGMENTS

There are many people to whom I am indebted for the production of this book. A few supplied photographs, some advised, many others loaned amplifiers, but every one made significant contributions. To all those mentioned here, I offer a heartfelt "thank you."

Jerry Blaha of Amp Crazy, Robert Stamps of the Amp Shop, Eva Andréasson, Richard Stoerger of Audio Design Associates, Dan Baccelliere and Valentino, Severin Browne, Julian Wright of Celestion, Mitch Colby, conrad-johnson, the Doyles, Alexander Dumble, Ritchie Fliegler, Bruce Gary, Aspen Pittman and Rick Benson of Groove Tubes, Guitar Center, Guitar Guitar, Brent and Patti Magnano of Guitar Oasis, Jimmy's Guitars, John Pegler of Guitars West, Bruce Gittleman of Guitarville, Chris Albano and Jon Eiche of Hal Leonard, Skip Henderson, Harry Joyce, Howard Leese, Scott Leibow, Greg Bayles of Make'n Music, Marshall and Korg USA, Jim Marshall, Mark Sampson and Rick Perrotta of Matchless, Steve Melkisethian, Scott Tubbs of MESA/Boogie, Norman Harris of Norm's Rare Guitars, Roger and Brett of R&B Rentals, Carl Nielsen of Rockbottom, Paul Thompson, Via Faruqi of Tube Technology, Peter and Jacob of UCI Photographic, and Scott Frankland of Wavestream Kinetics.

Michael Doyle, 1995

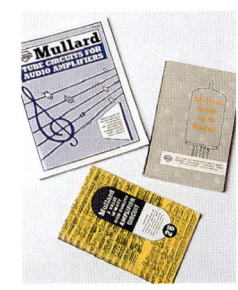

CONTENTS

Acknowledgments	3
Ampeg	6
Atwater Kent	8
Burns	10
conrad-johnson	11
Dumble	12
Fender	14
Gibson	24
Gretsch	30
Groove Tubes	32
Hiwatt	35
Magnatone	40
Marshall	42
Matchless	52
McIntosh	53
MESA/Boogie	54
Music Man	56
Olympus III	58
Orange	59
Rickenbacker	62
Selmer	64
Soldano	67
Tube Technology	68
Vox	71
Watkins	77
Wavestream	78
White	80

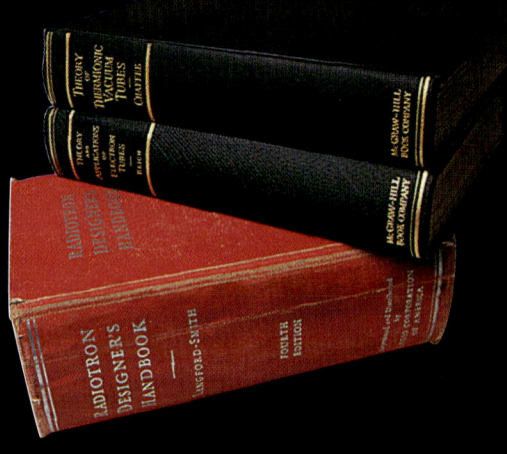

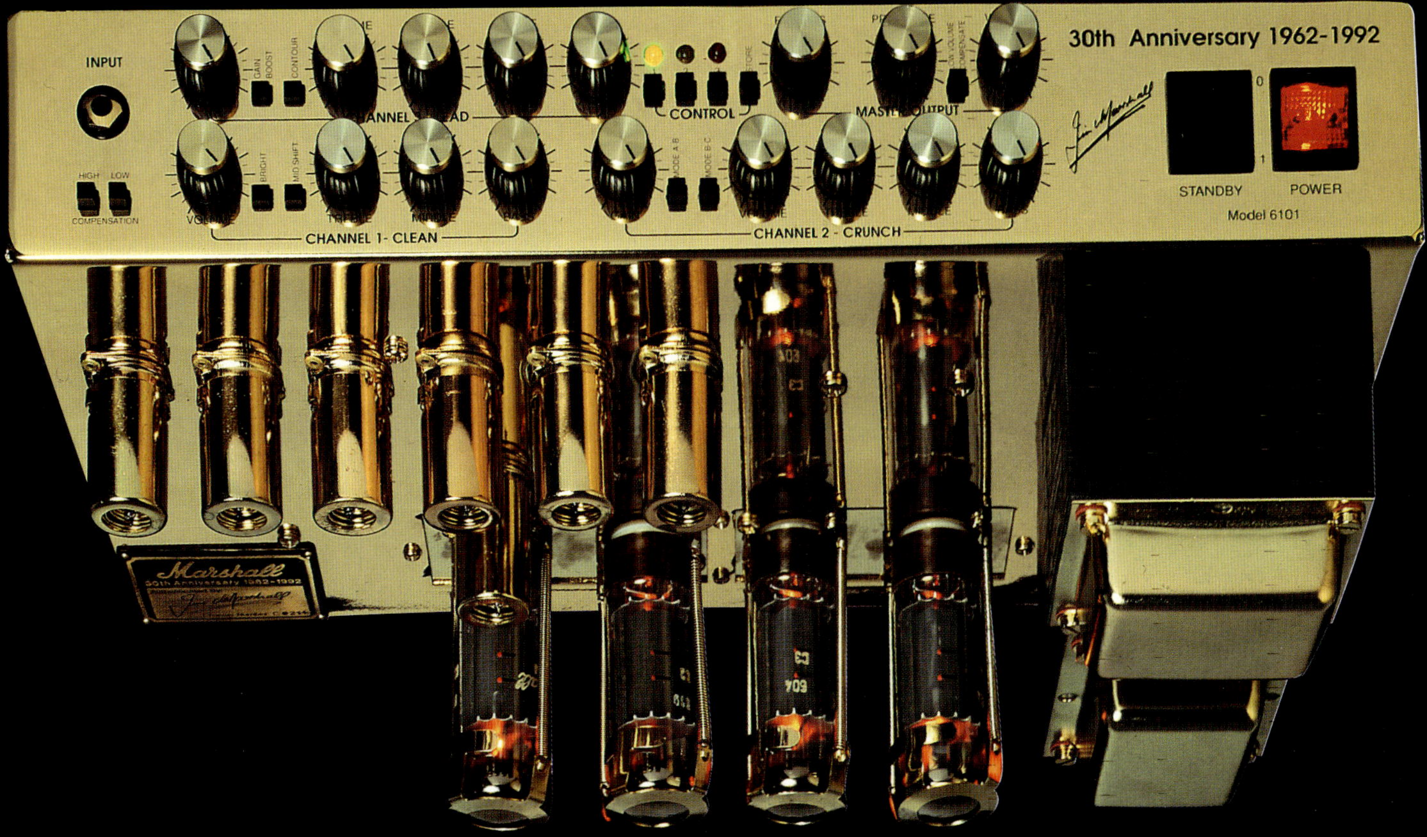

"I think the stroke of genius, really, was not his inventing the electric guitar, but inventing the amplifier to go with it."

Keith Richards, inducting Leo Fender into the Rock and Roll Hall of Fame, 1992

AMPEG

Everett Hull had a way with words—and a bass that was too quiet. By attaching a microphone to the end pin (peg) of a stand-up acoustic bass and plugging the mic cable into an amp, Hull created the "Amp Peg." His penchant for names spread to amplifiers, and, reflecting the "jet age" that had just arrived and the "space age" that was looming, he graced the bandstands of America with amps called Gemini, Jet, and Rocket.

It is Ampeg's contribution to bass amps, though, that has stood the test of time, beginning with the innovative Portaflex (*facing page, bottom*).

Covered in Blue Diamond Tolex and sparkle-weave grille cloth, with lashings of chrome and accordion inputs (and hard-to-find tubes), Ampegs had a look and a sound all their own.

Ampeg Rocket
GUITAR
ACCORDION
MIC
VOLUME
TONE

MODEL NO. R-12
SERIAL
Tube Location (from left to right)
6SL7 6SL7 6V6 6V6 5Y3
Ampeg Co.
54-06 Roosevelt Avenue
Woodside, New York

Jensen
CONCERT
DP-ALNICO-5

Jensen
CONCERT
DP-ALNICO-5
MADE IN U.S.A.

Ampeg
Ampeg Portaflex

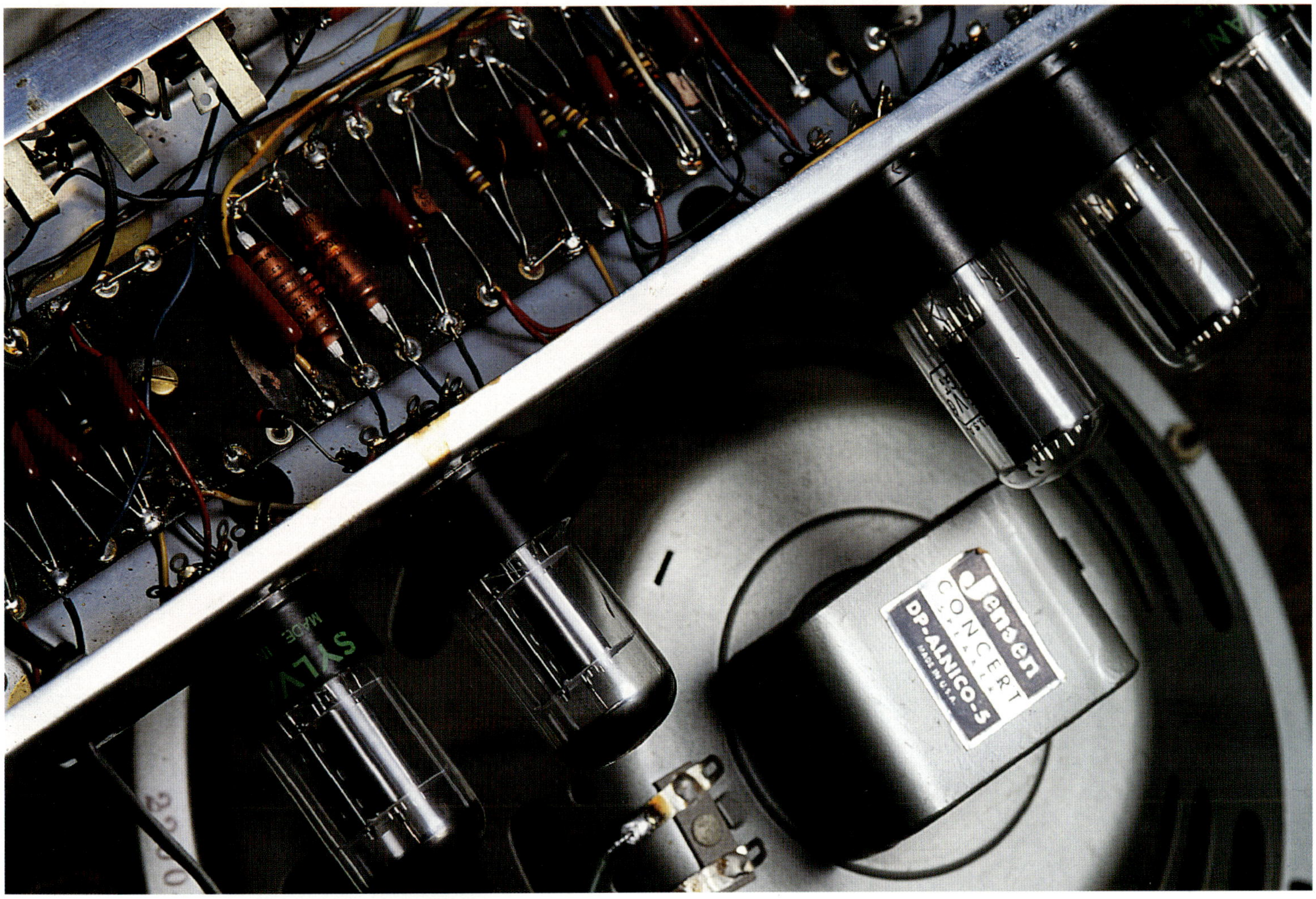

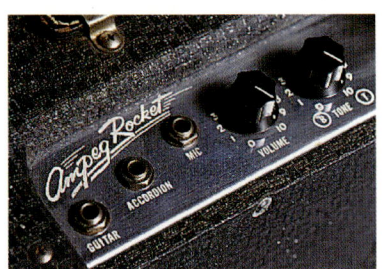

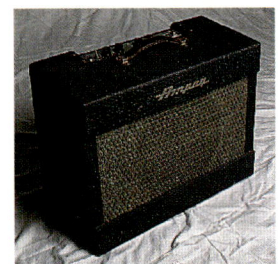

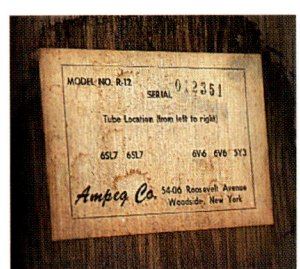

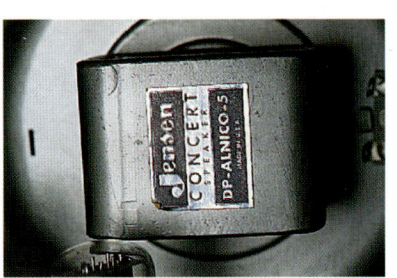

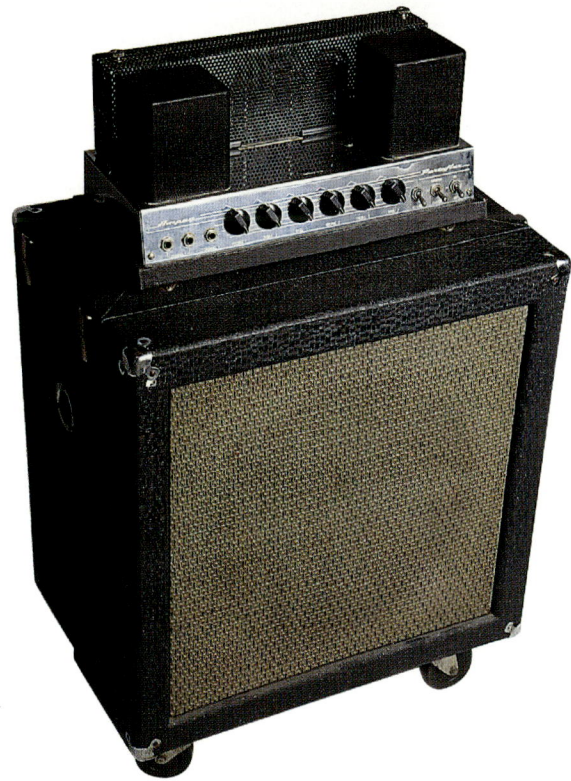

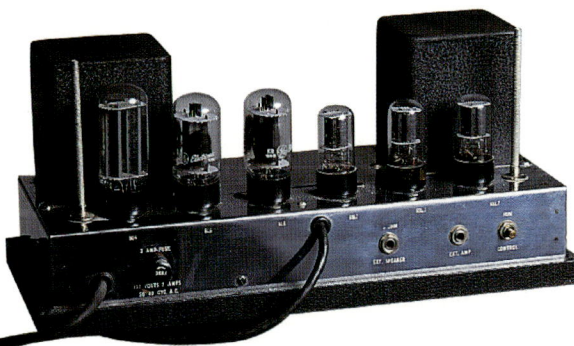

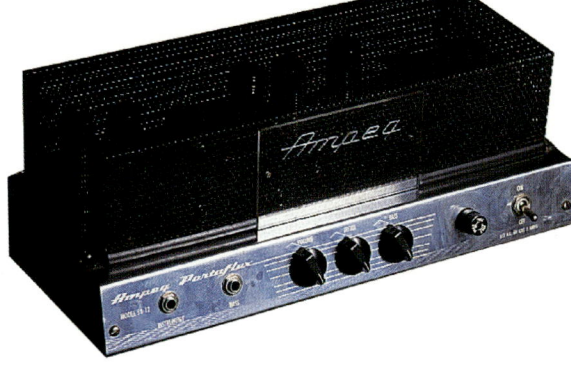

INSTRUCTIONS FOR

ATWATER KENT

OPEN MOUNTED STYLE

RADIO RECEIVING SETS

ATWATER KENT MANUFACTURING COMPANY
4700 WISSAHICKON AVENUE
PHILADELPHIA

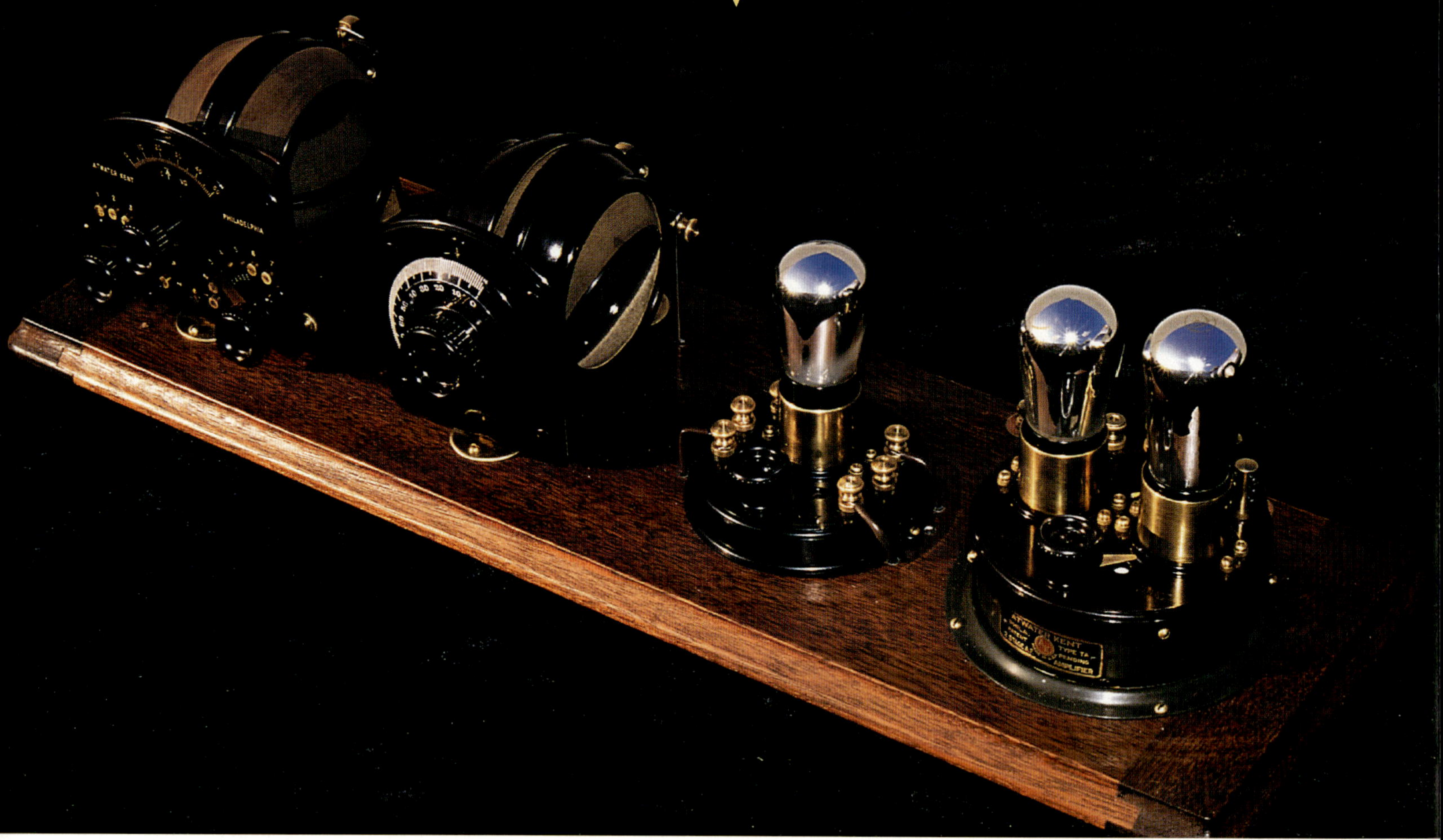

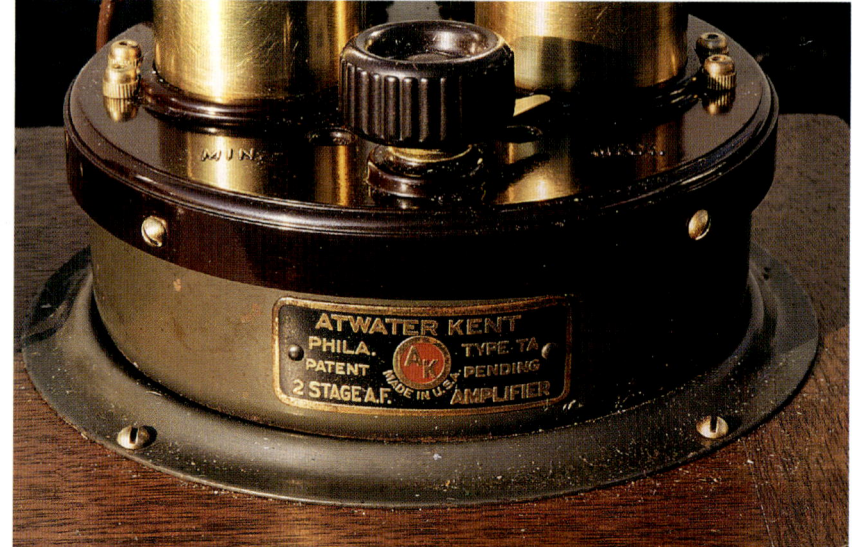

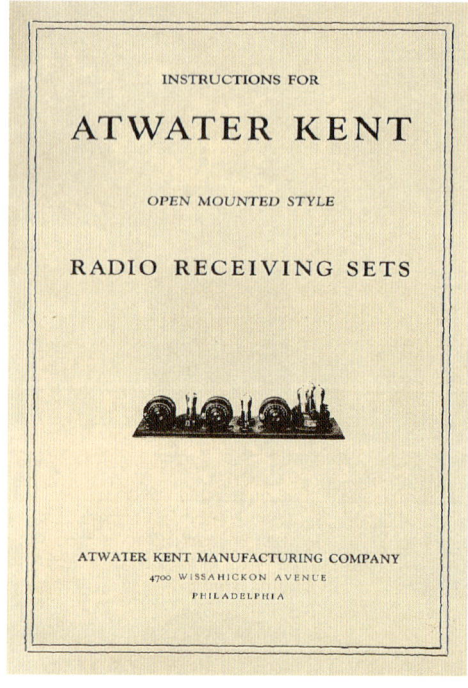

Musical instrument and audio amplifiers all have their roots in the radio sets of the early 1900s, and surely the finest of these are the Atwater Kents. Introduced in 1922, the "open mounted" (now commonly referred to as "breadboard") radios were stunning in their beauty. Note the ornate Bakelite dials, stained wood board, and brass tube sockets—prime examples of the "if you've got it, flaunt it" school of design. Unfortunately, this style wasn't in vogue for long, and by 1924 the enclosed "cabinet" style was in production—altogether more dull and dreary.

In 1929 Atwater Kent introduced a radical departure from the norm: the Model 57, which was manufactured from a metal frame. The princely retail price of $105, together with the undoubted complexities of making such a cabinet, ensure the rarity of these units today.

It's sad to report that the depression hit Atwater Kent hard, and by 1936 they were out of business.

"The radio tube is a marvelous device...."
RCA Radiotron Tube Manual, 1933

BURNS

In an era when "all transistor" was considered a virtue, Jim Burns offered the aspiring British musician a choice of some fine instruments, although in all honesty the Sonic 30 shown here probably wasn't one of them.

Legend has it that, upon his arrival in England, Jimi Hendrix was given Burns amplifiers to use, and so unhappy was he with them that, in order to substantiate his claim that they were unreliable, he threw them down the staircase!

Burns's main claim to fame was the use of their guitars by Hank Marvin of The Shadows; however, their amplifiers were almost inevitably overshadowed (!) by the Marshall and Vox products of the period. They do have a certain aesthetic appeal, though; the Sonic almost shines here in the reflected glory of the splendid Marvin guitar.

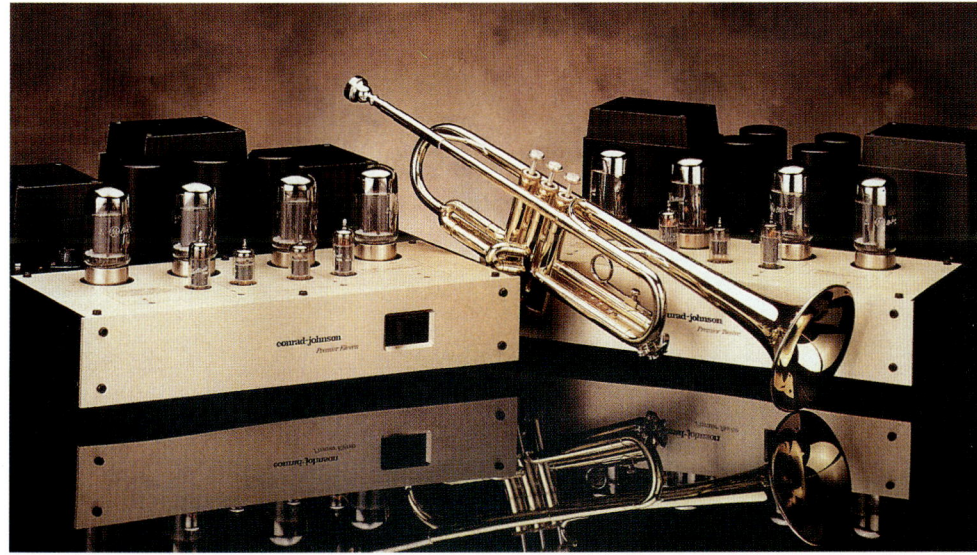

While tube guitar amplifiers managed to prevail against the onslaught of solid-state technology during the sixties and seventies, their hi-fi counterparts were not so lucky. However, a resurgence of interest in the last decade or so has seen the tube audio amplifier regain its position as the technology of choice in the hearts of the wealthy music lover.

So, for those who insist that the finest in musical reproduction should be accompanied by the warm glow of 14 tubes, or 28 in stereo, conrad-johnson offer you the monaural 275-watt Premier Eight (*above*) for a mere $15,000 a pair. For those who can only afford the Rolls but not the Royce as well, the 70/70-watt Premier Eleven and 140-watt mono Premier Twelve (*left*) should do nicely.

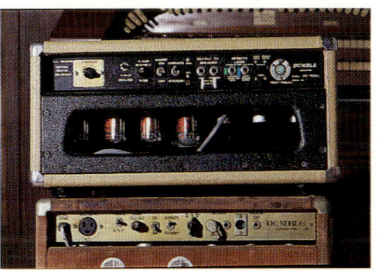

Some people nurture a reputation for eccentricity and mystique. Alexander (previously Howard) Dumble *can't help* being eccentric and mysterious. Living in an old oriental house, which adjoins an abbey complete with its own chapel and dungeon, Dumble builds some of the world's most coveted and certainly most expensive guitar amplifiers at the current rate of two a year.

Wooden circuit boards, unique tone circuitry such as the "starve" switch, and custom suede coverings are all Dumble trademarks.

Known on occasion to determine the suitability of potential friends by their ability to *mentally* calculate the square root of two, Dumble rides a Harley, pans for gold, and builds his own pistols and rifles—just like everybody else.

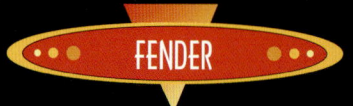

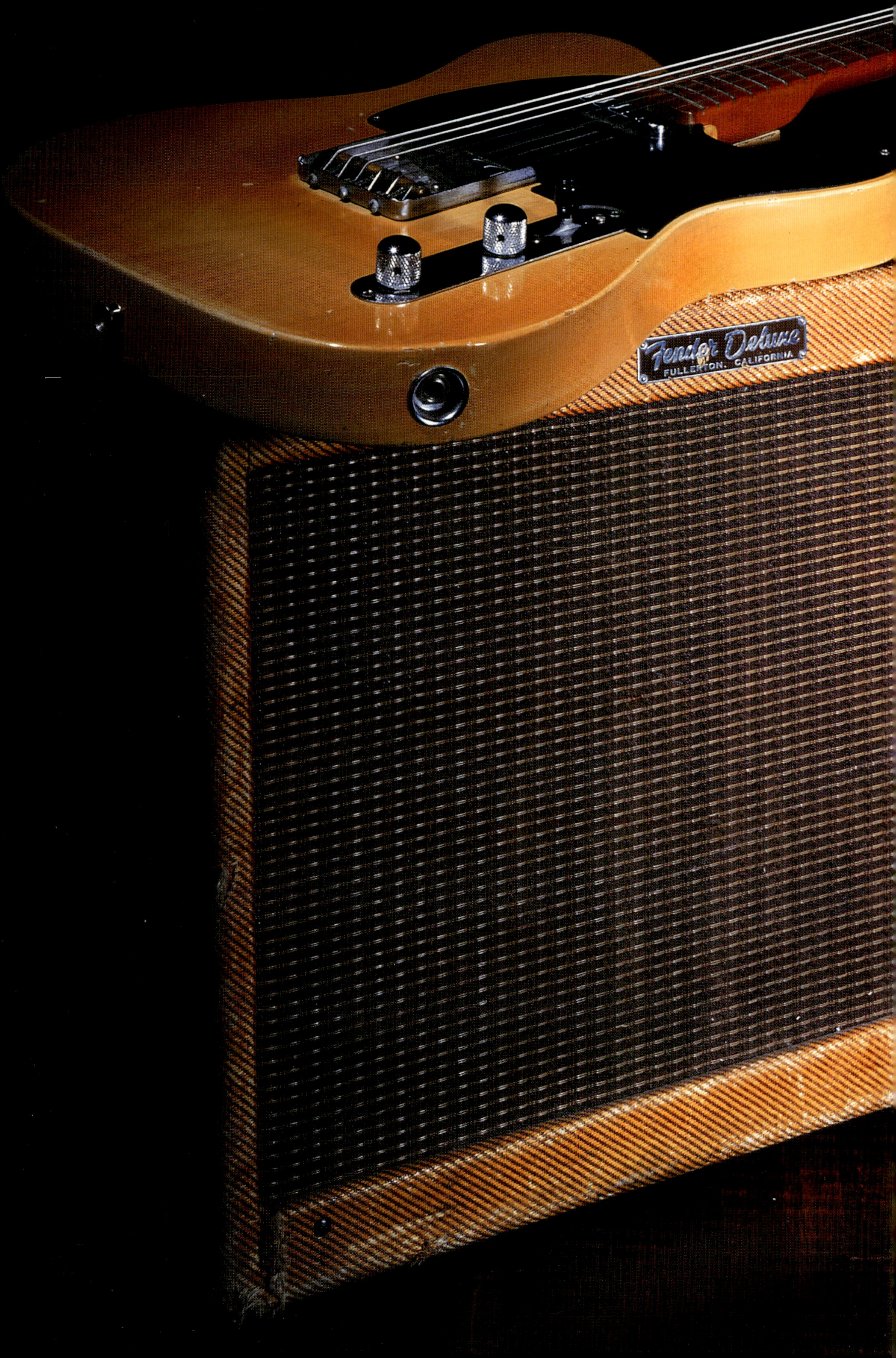
Fender Deluxe
FULLERTON, CALIFORNIA

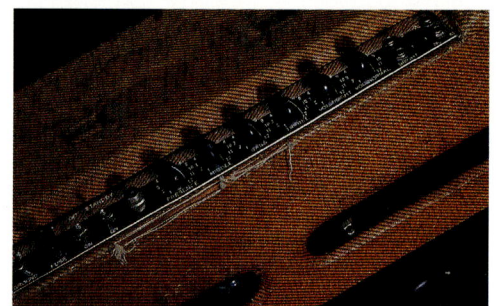

Where would we be without Leo Fender? Although he didn't exactly invent either the electric guitar or the amplifier—Keith Richards notwithstanding—his designs loom larger than any others. One of the most enduring and influential of these was the Deluxe amp (*above, below, left, and pages 14–15*).

With a lineage that traces back to 1946, the tweed-covered "narrow-panel" was introduced in 1955 and survived until the release of the brown Tolex-covered version in 1961.

The Bassman model 5F6-A (*right and page 17*), announced in 1958, is considered by many in retrospect as the greatest of all Fender *guitar* amps. It was curiously discontinued in 1960, perhaps because Leo thought it distorted too much.

"Avoid playing the amplifier at a volume setting high enough to produce a distorted sound through the speaker."

Fender Electric Guitar Course, 1966

Although Hendrix was probably raping his amplifiers and burning his guitar as this statement was written, there were nevertheless many musicians trying to "keep it clean." They needed look no farther than the Tremolux. Resplendent in its white vinyl and cool chrome legs, it was the epitome of early sixties chic.

With its distinctive "V" front, lightning-bolt logo, chrome control panel, knobs that go to 12, twin speakers (an industry first), and production history that occupied only the year 1947, it's hardly surprising that the original Dual Professional (*pages 19–21*) is the most coveted of all Fender amps.

Curious design details abound. The output transformers are mounted on each of the speakers, and the vertical chrome strip on the front actually holds the two baffle boards together. In 1948 the name was changed to Super and the lightning bolt was gone, and by 1953 so was the angled cabinet. Our loss.

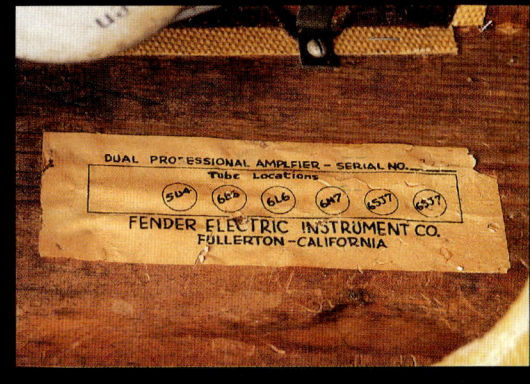

"Tube mania: there is no cure."
Fender schematic, 1990

Justifiably more famous for their guitars than for their amplifiers, Gibson have nevertheless built some fine-looking amps in their time. With its distinctive striped tweed covering, the EH-185 is the most attractive of all. It was called the 185 because in 1940 it could be bought for $185 complete with a six-string steel guitar; coincidentally, its power consumption is 185 watts.

The ported 1x12 cabinet has a flip-top compartment that contains the removable seven-tube amplifier. This feature was expanded on by Ampeg many years later with the Portaflex.

The guitar is a 1950 ES-5.

As an extension of their successful guitar-making partnership with Les Paul, Gibson produced a matching line of amplifiers. The Les Paul GA-40 and Les Paul Junior shown here were two of the most popular models, although over the years they were offered in a variety of different cabinet styles.

The Junior was offered as part of an outfit with the Junior guitar and was a three-tube design with an oval Rola speaker, producing 4 watts. The GA-40 was a serious piece of work and was described in one of the catalogs as being a "top quality high gain" design! The seven tubes gave 15 watts through the 12" Jensen speaker, and in 1958 the amp cost $199. For comparison, a Flying V was $247.

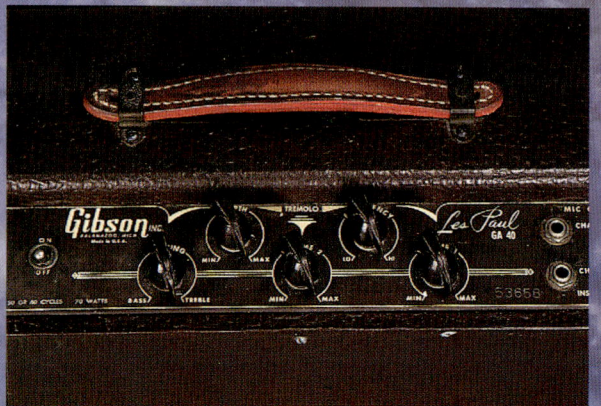

Gibson introduced the 79 RV stereo amplifier around 1960 to go with their line of stereo guitars. Reverb, two 10" speakers, stereo or mono outputs, eight tubes yielding 30 watts, and inputs for a turntable (!) made this quite a comprehensive little package. The 79 RVT had all this and tremolo too for $395.

In the large photo on these two pages (and on the right above) is a fantastic example of a 1952 GA-50. Note the as-new metal can-type RCA preamp tubes and the grey painted output tubes, which produce about 25 watts.

The other amp is a GA-30 from 1951. Slightly lower powered, it still shares a similar speaker combination of an 8" and a 12".

GRETSCH

Better known for their drums and guitars, Gretsch nevertheless did make a few amplifiers in the fifties, called Electromatics. Some incorporated such noteworthy features as a built-in tweeter, "wrap-around grille," and polka dot covering to match their guitar cases.

The larger amplifier shown on this page is an Electromatic Deluxe, with a 15" speaker and seven tubes, producing up to 25 watts. For a $15 premium a singing cowboy could replace the polka dots with a "Western finish" and tooled leather trim. The 1957 Rambler guitar is leaning against an Electromatic Standard, which is notable for a combined bass and treble control.

Shown with the refinished blue 6118 Double Anniversary on this page is my favorite of the trio, a 1949 Electromatic Standard. It's the details that impress me: a cast, chrome-plated logo; silver speaker cloth with matching stripe; and a tube chassis that's perfectly symmetrical, with an 8" Rola speaker at its head. A 6-watt jewel.

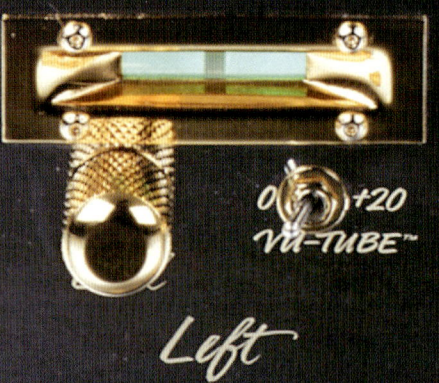

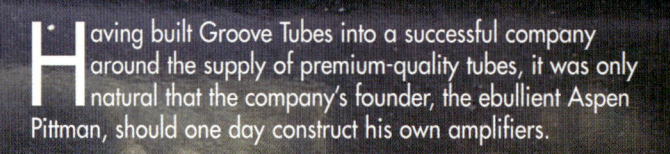

Having built Groove Tubes into a successful company around the supply of premium-quality tubes, it was only natural that the company's founder, the ebullient Aspen Pittman, should one day construct his own amplifiers.

The Dual 75s Limited combines modern construction with the finest in retro chic styling. Dripping in chrome and brass plating, with two "Vu-Tubes," it was designed primarily by the great steel guitar player Red Rhodes, who, sadly, died in August 1995. Beyond doubt one of the best-looking tube amps ever made.

Hiwatt was the brainchild of an Englishman named Dave Reeves. Originally employed by Sound City, Dave came to the conclusion that he could build better amps (it wouldn't have been hard). As luck would have it, he came across a government-certified wirer named Harry Joyce, who apparently was a perfectionist and who agreed to wire a limited quantity of chassis for Dave's new amp. Introduced in 1964, the first Hiwatt amp, with top-mounted chassis, bore a striking resemblance to the first Park 45s (built by Marshall). Ditching this style of chassis for being too complex (as did Marshall), Reeves replaced it with the more traditionally styled 50-watt head shown at upper left, with the stylish black-on-gold plexi logo.

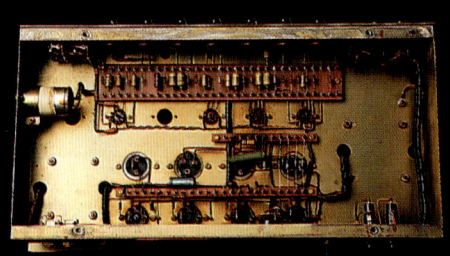

By about 1967 Hiwatt finally found their look and stuck to it for years to come. Made famous through endorsement by Pete Townshend, the black and white monsters with their ripping rhythm tone seemed to have a healthy future. But it was not to be. Production was always limited due to the high standards of construction, and this problem was compounded by the location of the "factory" at the back of a suburban bakery! Nevertheless the company survived until the eighties, when Dave Reeves tragically died. The name has been passed around since then with varying degrees of success.

NORMAL

BRIGHT

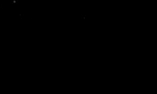

NORMAL VOLUME

BRIGHT VOLUME

BASS

TREBL

JOYCE

While the Hiwatt name is owned by others, it appears as though Harry Joyce is the spiritual heir to the throne, as evidenced by the introduction in 1993 of the Joyce amplifier.

The same high standard of construction is still clear to see, as are the Partridge transformers and EL34s. Better yet, the colorful internal wiring layout will forever remind me of the London Underground map!

100

STANDBY   MAINS

MIDDLE   PRESENCE   MASTER VOLUME   ——— ON ———

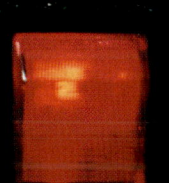

40

> **"Big as all outdoors—the vastness of the sky combined with Magnatone's big 'V' Electronic True Vibrato."**
>
> *Magnatone catalog, 1957*

Fender, amongst others, *claimed* to build amps with vibrato, but in 1957 only Magnatone offered true vibrato, instead of the more common tremolo—an important distinction.

Immediately identifiable by the Cadillac-style "Big 'V' " on the front grille, Magnatones offered the patented vibrato circuitry on models such as the Custom 260 shown here, which featured two 12" speakers and eight tubes for 35 watts. At $295 it was definitely aimed at the pro end of the market, and none other than Buddy Holly used Magnatones to great effect, especially, I would imagine, on his classic "Words of Love."

Yet Magnatone amplifiers were originally small, plastic-covered tube units sold with Hawaiian steel guitars between 1947 and '55. Fine though I'm sure they were for this application, plug a Les Paul into one of these little beauties and prepare to be amazed. Best of all, they were offered with surf scenes on the grille cloth; an example of one of these can be seen on page 3.

Although the tweed Bassman was discontinued by Fender themselves in 1960, its virtues were not lost on either Jim Marshall or his customers, and its circuit layout provided the inspiration when he set about building his first amplifier in 1962. Significantly, he made some fundamental changes when he resurrected Leo's masterpiece, including different tubes and transformers, higher output impedance, four 12" Celestions, and a head instead of a combo design. These principles have formed the blueprint of rock 'n' roll amplifier design ever since.

Pictured here is just about every variation on the early Marshall theme, with an unnumbered prototype in the top left, leading to the last of the breed at bottom right—the 1966 JTM 45. The chassis in front is serial number 13 and probably dates from late 1962.

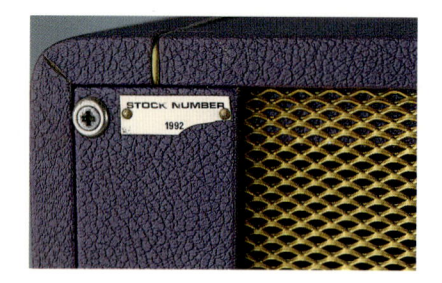

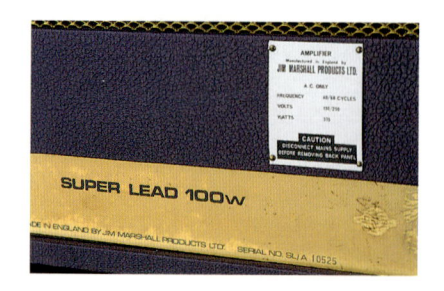

By 1966 Marshall had invented the 100-watt stack and were on a heady roll. Custom colors became part of their marketing arsenal, and "Royal Purple" was first offered in the 1968 catalog. Other colors available included blue, brown, yellow, red, white, and orange. Original, unrestored examples are exceedingly rare and particularly valuable.

In 1994 Marshall went back to their roots and reissued this ultimate collector's fantasy. Standing seven feet high with its extra-tall "Hendrix" bottom cabinet, basket-weave grille cloth, plexi panels, EL34s, and purple vinyl, it was a retro hit. It's an impressive sight today, but by the standards of 1966 it must have been absolutely terrifying.

The 9200 was launched in 1993 as Marshall's top-of-the-line power amp. It delivers 100 watts a side in true dual monoblock style, with a perspex window that offers the lucky guitarist a view of the eight 5881s glowing warmly inside. It's shown here atop the matching JMP-1 preamp.

Channel B
STANDBY
POWER

B

Bass
Middle
Shift Treble
Clean 1 Presence
Clean 2 Effect
 Map
 Channel

Power
Made in England

47

Model 6100

To celebrate their 30th Anniversary in 1992, Marshall decided to flex their stack muscles and build a limited run of 800 EL34-equipped three-channel heads in blue vinyl. If it was metal, it was brass plated, including the tube retainers. Known as the 6100LE, it was also available in combo form, but produced in even smaller quantities. The combo chassis is even more spectacular than the head, and is pictured on pages 5 and 50.

50

Here's a transition JTM 45, with an even rarer white-front 4x12 cabinet. The speakers used were the now-legendary Celestion G12 Alnicos, which were first manufactured around 1936! Updated with heavy-duty construction and painted blue for the Vox AC30, the standard silver version known as the T.652 was the first speaker used by Marshall in their 4x12s. Note the yellow foam used as acoustic insulation in these early cabs.

MATCHLESS

The Matchless Amplifier Company was born of a love for "the classic English tube amps of the early sixties," specifically the Vox AC30. Far from being blatant copies, however, Matchless amps encompass many of the requirements of the modern player, with a quality of construction that justifies a lifetime warranty.

Their flagship amplifier is the vintage radio-inspired D/C-30 Exotic. The beautiful 1x12 or 2x12 cabinet, crafted from exotic wood such as wenge, maple, koa, padauk, or walnut, as in this example, is obviously its most distinguishing feature. But the lavish detail also extends to the chrome-plated chassis and gold-plated transformer covers and tube mounts.

Four EL84s and a pair of 5V4G rectifiers ensure that it sounds as good as it looks, while production of one a month at $5,600 each ensures exclusivity.

McINTOSH

Much of the interest in today's tube hi-fi amplifiers is as a consequence of the outstanding quality of the original tube amps of the fifties. And perhaps the most coveted of all those great names of the past is McIntosh.

Their reputation for sonic performance is legendary, and much of that is due to their unique and patented "Unity Coupled" transformer design. Such is the popularity of original examples such as this MC-60 (a 60-watt monoblock) that the company have reissued in limited quantities of some of their earlier designs, albeit at enormous expense.

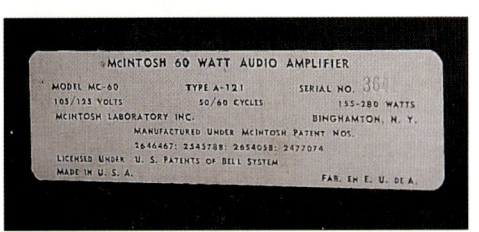

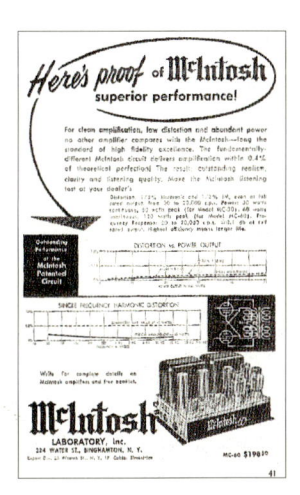

ORANGE CHANNEL GAIN
CLEAN RHYTHM
CHANNEL CLONING
CHANNEL SELECT
EXTERNAL SWITCHING
SPEAKERS
8 OHM
4 OHM
LEVEL
SLAVE OUT
HANDTESTED IN AMPLIFIERS & COMPUTER MATCHED
TRIPLE
Rectifier
SOLO HEAD

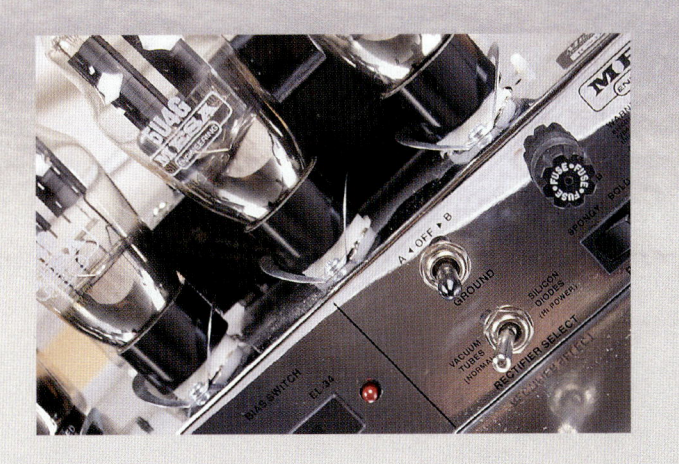

The first time I saw this I did a double (or should that be triple) take, not quite believing what my eyes were telling me. Triple rectifiers! And curvaceous 5U4Gs at that! Lending a "tubular" balance to the aesthetics, there are another six Coke-bottle 6L6GCs all stretched out in a neat row like they were waiting for a bus! I've always admired Boogies for their attention to detail and tactile qualities, but this chrome-plated colossus takes those values to an even higher level.

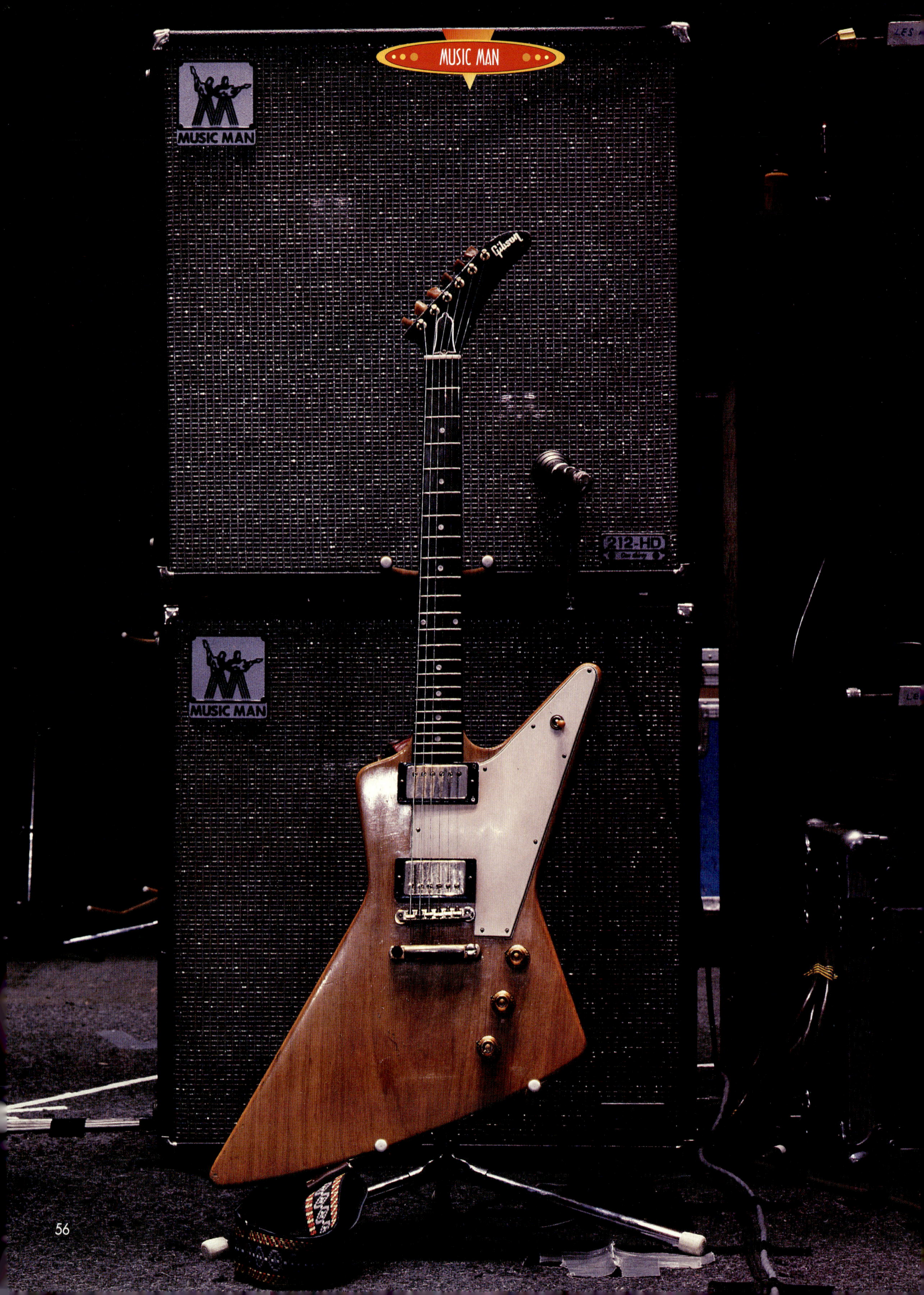
MUSIC MAN
MUSIC MAN
Gibson
212-HD

When Music Man began building hybrid amplifiers—that is, those incorporating both solid-state and tube circuitry—in the mid seventies, the idea wasn't new. But the company were nevertheless one of the first to popularize the technique.

They built a wide range of models, from small 1x12 and 2x10 combos to stacks rated as high as 150 watts. While they clearly made 4x12 cabinets, the cabs on page 56, partially obscured by Eric Clapton's Explorer, are actually 2x12s.

Highly regarded in the seventies, they have lost some of their lustre over the years and are now available at bargain basement prices.

Music Man discovers MUSIC MAN

OLYMPUS III

Manufactured by Audio Design Associates, the Olympus III is in a league of its own in both construction and application. It's built from *solid brass*, which is then hand-polished and plated in *24-karat gold*! Inside the golden cage are 15 tubes, plus a further six vacuum display tubes for visual monitoring of each channel through the window on the front panel. So what is it? It's the world's one and only *tube* Dolby Surround Sound Controller.

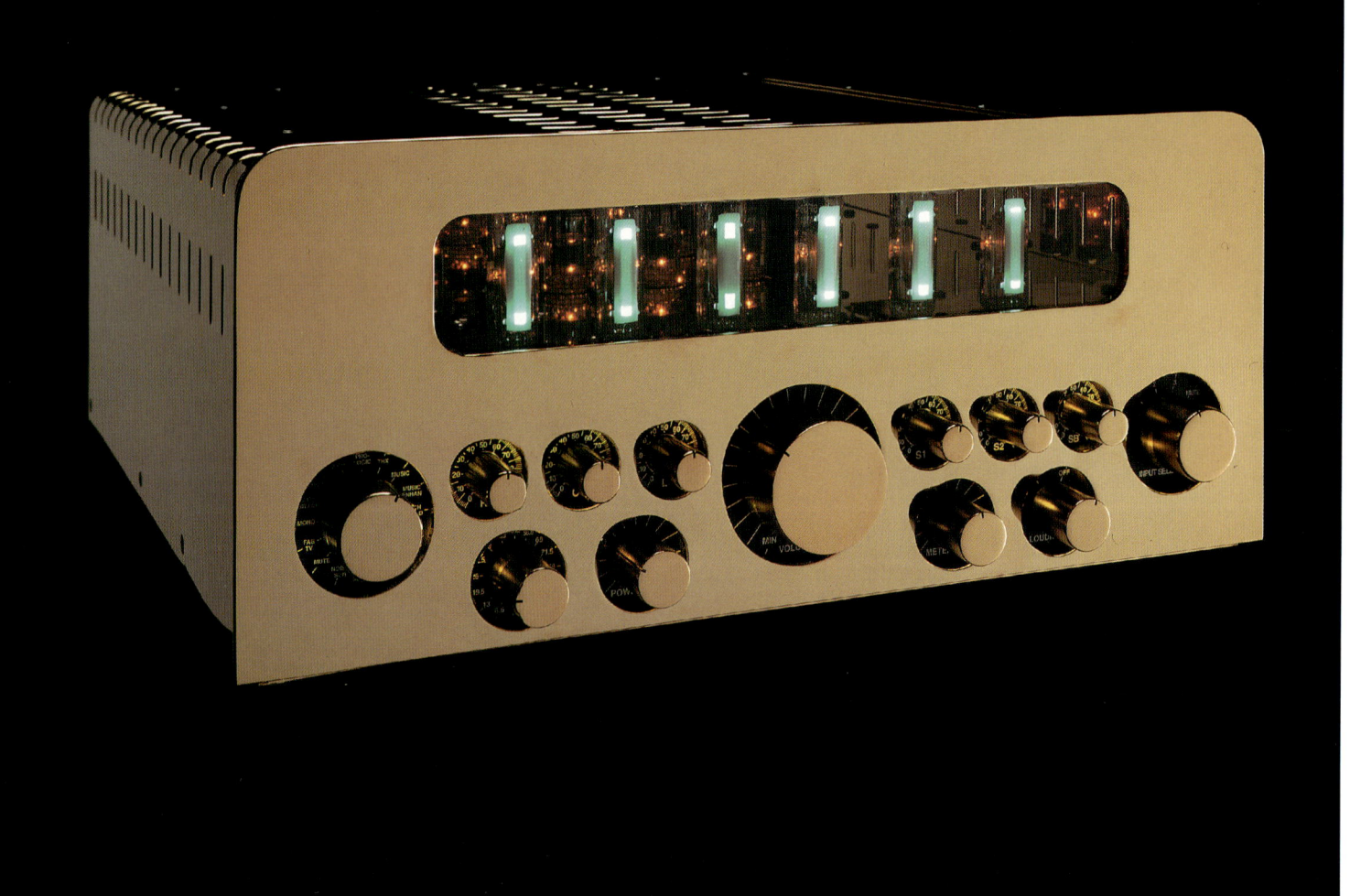

ORANGE

Orange Musical Industries, born in London in September 1968, had lofty ambitions. They aimed to combine a "recording studio, record company and labels, publishers, agency, management, retail and wholesale outlets and, last but not least…amplifiers."

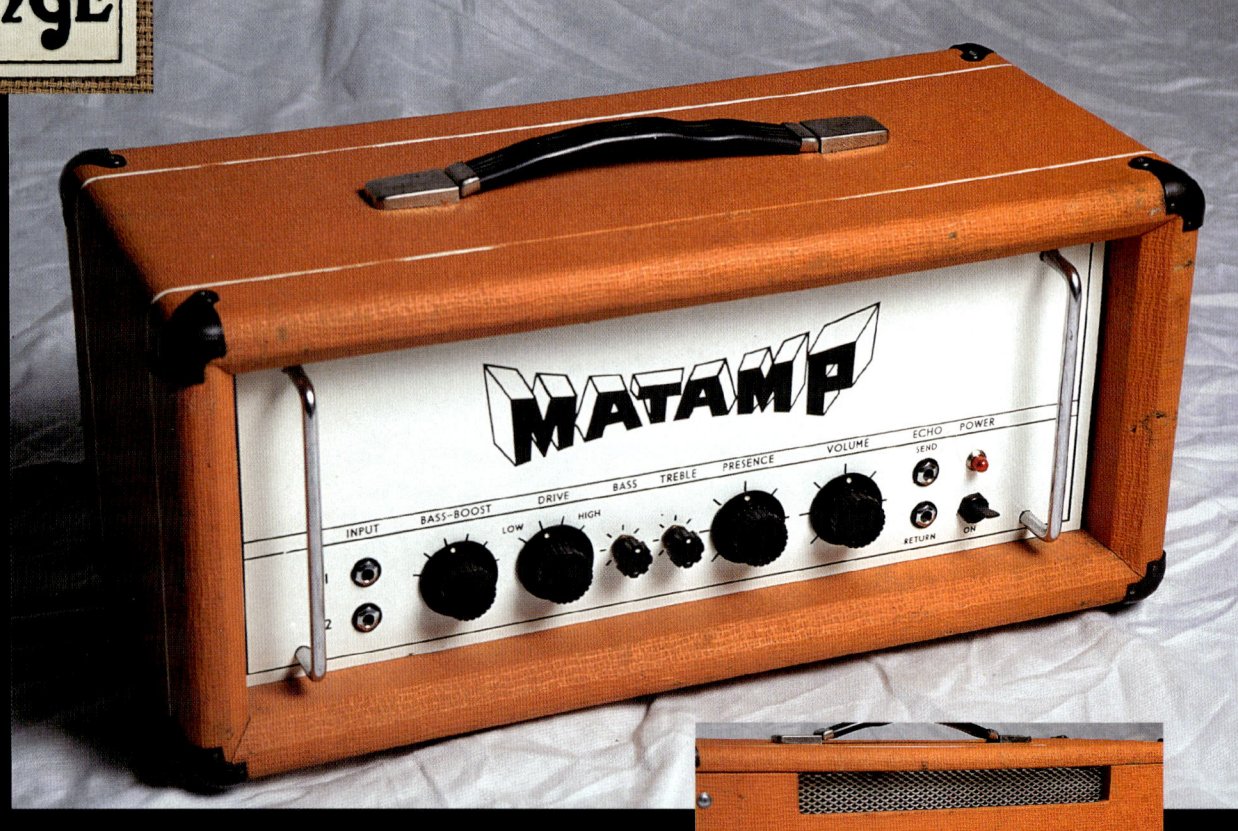

ORANGE

With construction details inspired by Hiwatt, such as Partridge transformers and mil spec wiring, Orange added their own unique touches, like orange enamel-painted chassis and "chrome roll-over cradle bars"! Best of all were the hieroglyphics on the control panels—quite brilliant.

Both tube and solid-state designs were offered, the largest tube amp being a 200-watt slave, while they did also make "possibly the largest guitar amplifier cabinet in the world, weighing half a ton and incorporating 24 12" speakers"! The Matamp was basically the same product, but with a different distribution arrangement for the North of England.

They were quick to point out that the amplifier passed through "56 check points before the final Gold Seal of Approval is affixed." But they didn't stop there, oh no! They then had to affix the "world famous British Export Seal of Approval." You also got an owner's log book. But none of this altered the fact that not only didn't they sound that great, but they weren't particularly reliable, either.

Unfortunately, the best thing about Orange wasn't their amps, but their catalogs. So bizarre that they bordered on science fiction in some ways and Shakespearean prose in others, their vocabulary was as colorful as the amps they tried to sell. Here is an excerpt from their explanation of the "knights of old" coat of arms as shown in the middle of the cabinet pictured here:

"Central in the design is the lush foliage springing from the barrel of plenty. Symbolic of the wide range of quality equipment under the Orange label. The god of nature and hypnotic music stands to the left,—and not to forget Orange's important position in the British music industry—Britannia is on the right....At the bottom of the emblem, emblazoned across the musical stave, is Orange's world famous motto 'Voice of the World.'"

Well, now you know.

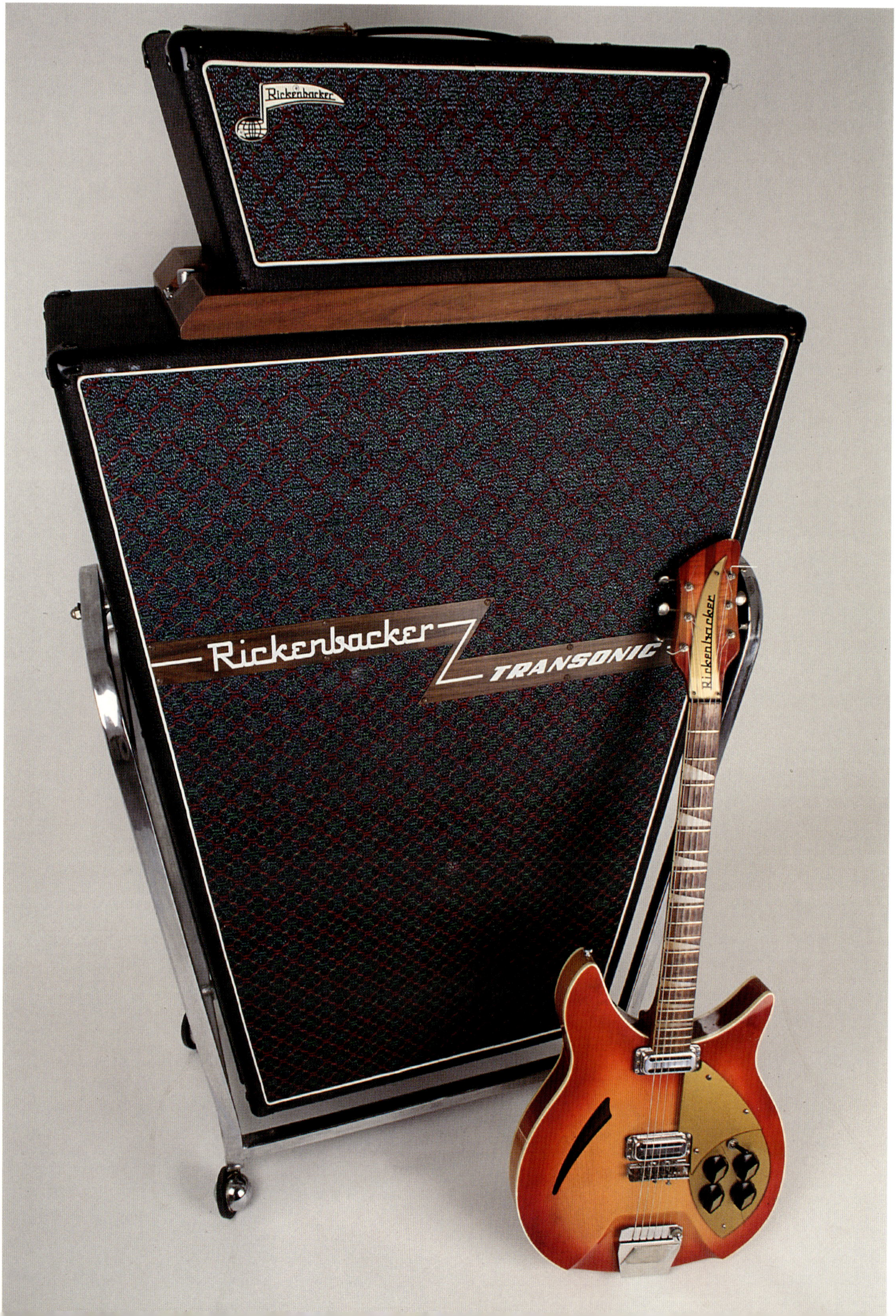
Rickenbacker
Rickenbacker
TRANSONIC
Rickenbacker

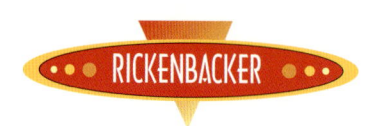

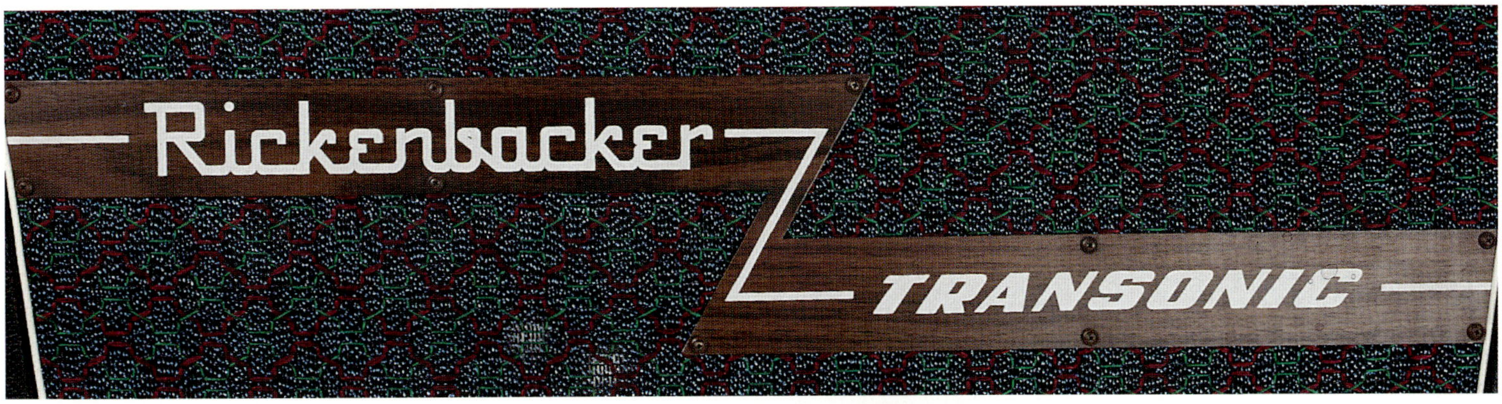

Rarely was an amp more appropriately named than the Transonic by Rickenbacker. Shaped like an inverted rocket, with a chrome tilt stand as its launching pad, it was announced in the late sixties in two different versions: the Series 100, with 200 watts of peak power, and the Series 200, with 350 watts peak.

Yet despite being used by Led Zeppelin, Jeff Beck, and The Doors, it wasn't a commercial success. This was probably due to its solid-state design, which, however, included built-in fuzz-tone, reverb, and tremolo effects. There was also an "overload" meter in case you had gone deaf and weren't aware of the distortion! Although the Transonic was available in a few different variations, this particular model has a 2x15 cabinet. By the way, the guitar is a 1959 Rickenbacker Capri.

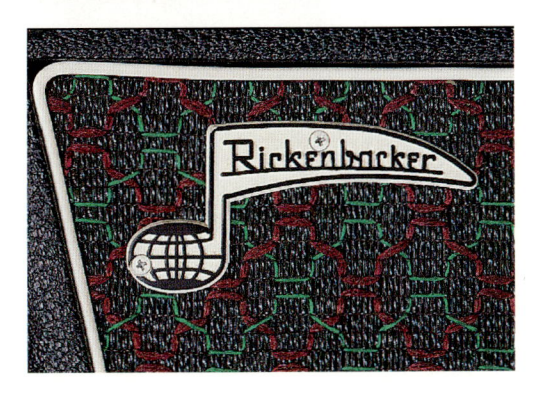

SELMER

Of all the British amplifier companies from the sixties who dropped the business ball—Orange, Hiwatt, Vox—Selmer's lack of success remains the biggest mystery. Consider the facts: All-tube designs using EL34s, KT88s, and GZ34s, foot-switchable tremolo *and* reverb with patented "blinking eye," 2x12 or *4x12* combos, and of course the incomparably stylish "duo-tone croc-o-hide" coverings. They even made 100-watt tube amps complete with EL34s and *two* GZ34s—rack-mountable—as well as offering switchable mono or stereo outputs—in 1964!

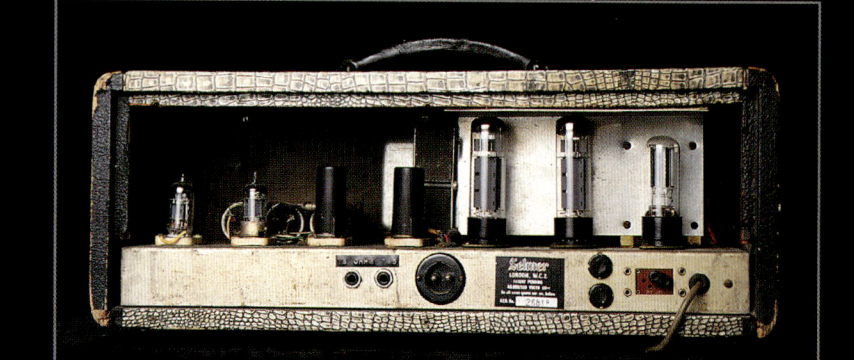

Shown here is the distinctive, all-purpose Treble-n-Bass Fifty, which, like the other Selmers pictured, dates from around 1964. The Selmer combos were popular, especially the Zodiac Twin 50 shown on the following page. Also shown is the Truvoice Bassmaster piggyback, which ordinarily would have used the "croc-o-hide," but this example features a rare two-tone blue covering.

Within a year, though, Selmer had redesigned the cosmetics of their tube amps to be spectacularly ugly, at the same time introducing solid-state designs, and I can only presume that that was the beginning of the end for this potentially great amplifier company.

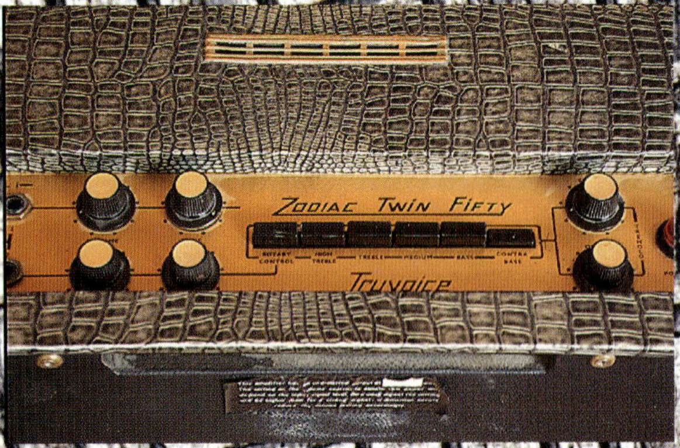
Zodiac Twin Fifty
Truvoice

Selmer
Truvoice
Bassmaster
Selmer

NEW Bassmaster 50
Selmer

Selmer

Mike Soldano does a pretty good job of being the all-American rock and roller: he drives hot rods and plays rock guitar. Better than that, he builds his *own* hot rods and makes his *own* amplifiers. His car license plate even says "AMP MAN."

Back in 1988 Soldano practically invented high-gain amplification when he announced the SLO-100. An interesting combination of American and British amp design ideas, together with plenty of his own, the Super Lead Overdrive (SLO) had knobs that went to 11, with a distortion to match.

While purple seems to have become Soldano's preferred color scheme, snakeskin I think is a more accurate representation of the amp's true character. I hope that Valentino, a particularly intimidating red-tailed boa constrictor, agrees.

TUBE TECHNOLOGY

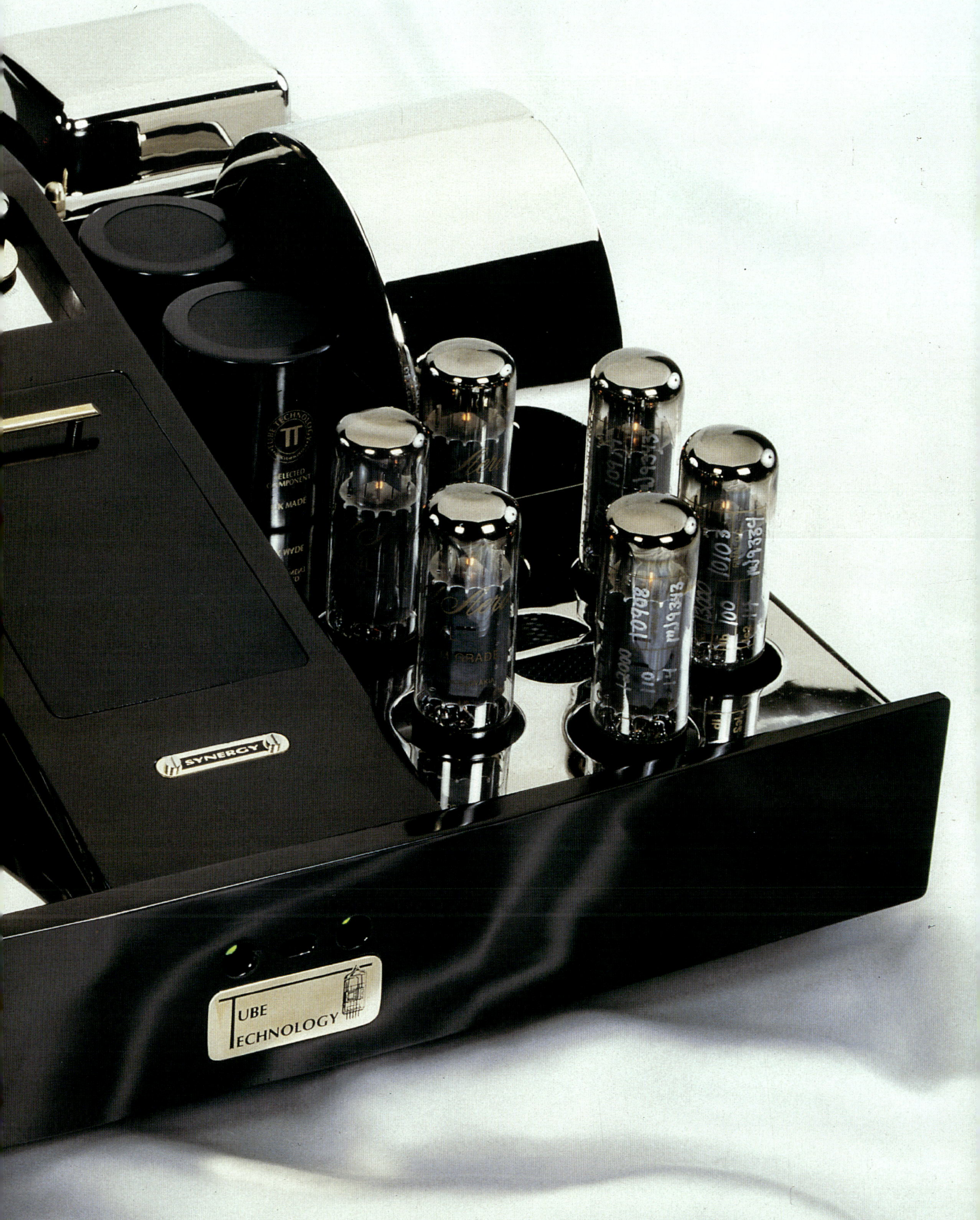
SYNERGY
TUBE
TECHNOLOGY

If any one amplifier sums up what this book is all about, then the Tube Technology Synergy (*above and pages 68–69*) is it. It leaves me speechless.

The more compact Unisis (*below*) is an integrated 30-watt-per-channel stereo amplifier, with EL84s in the output section. As a more economical alternative to the Synergy, it has neither the remote control facility nor the on-board tube calibration and timer facilities of its more elaborate brother. Each, however, is a stunning example of the tube amp art.

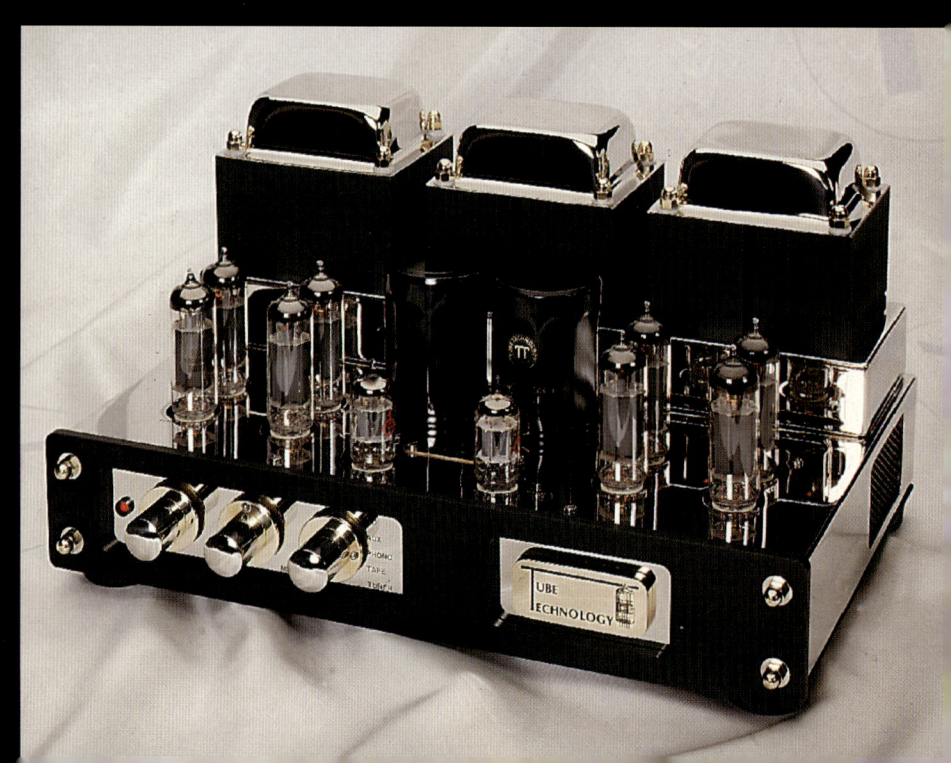

VOX

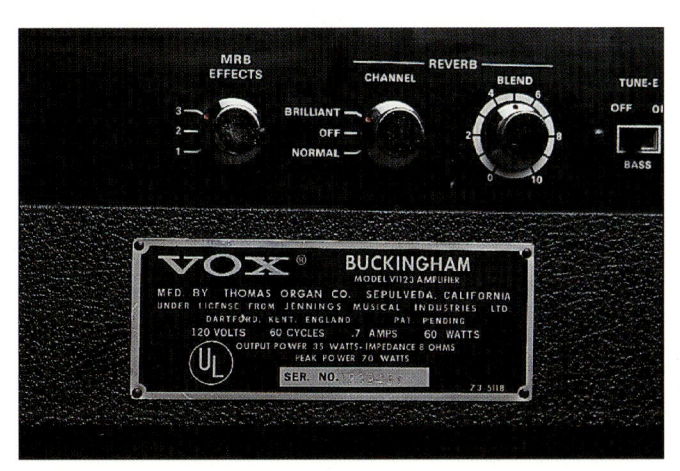

Riding the crest of a tidal wave in popularity thanks to The Beatles' use of the AC30, Vox launched a series of solid-state amplifiers that broke the company almost as quickly as it had been made.

The Buckingham shown here and on the next two pages was made in America and featured a "Remote Controlled Distortion Booster," tremolo, reverb, Top Boost, and a tuner—not bad for $475 in 1966. It also had a knob labelled "Tone-X" and a Standby switch. This last feature was particularly curious, since it was a solid-state amp!

Solid-state Voxes were notoriously unreliable, as well as being almost impossible to service. To top it off, they sounded nothing like the tube AC30s used by The Beatles. Consequently, the Vox reputation didn't survive long. If only looks were everything.

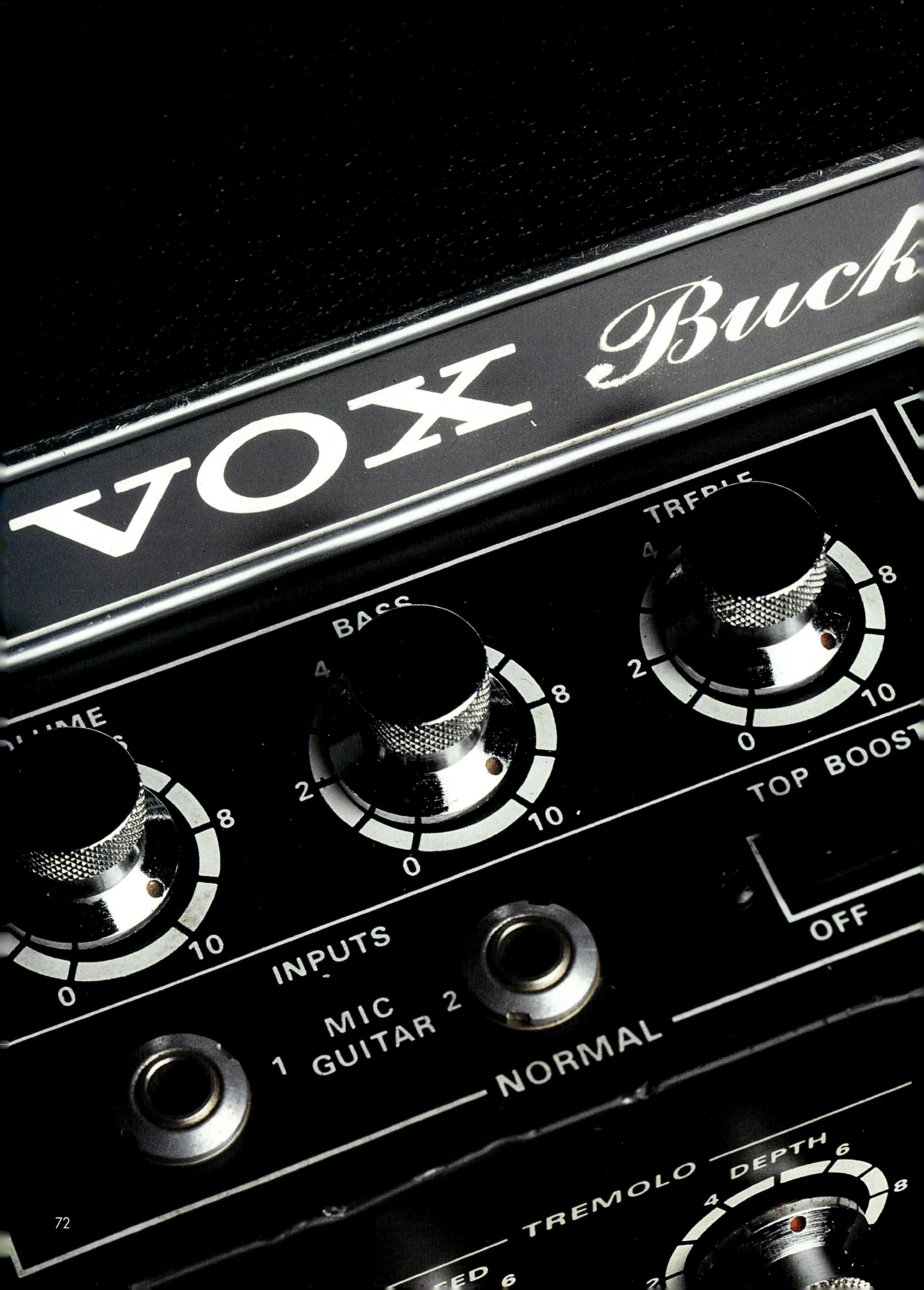
VOX
TOP BOOST
OFF
INPUTS
MIC
1 GUITAR 2
NORMAL
TREMOLO
DEPTH

ingham

TREBLE

BASS

VOLUME

MIDDLE B

INPUTS

BRILLIANT

ON

MRB EFFECTS

3
2
1

PEDAL

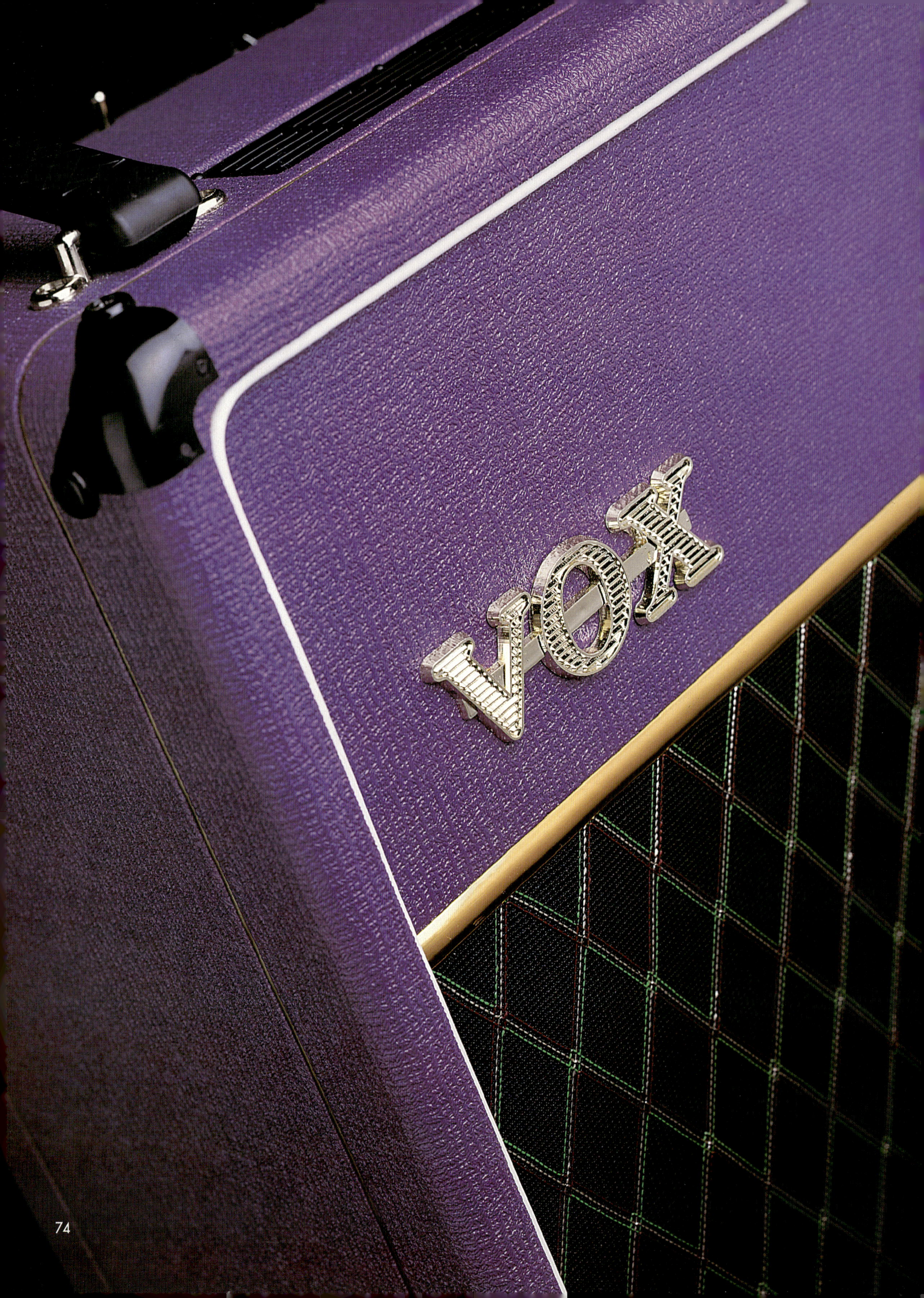
VOX

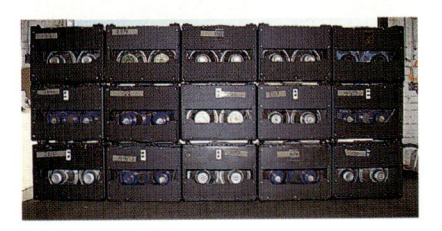

The first truly great British guitar amplifier was the Vox AC30, announced in 1959. The circuit, which had its roots in the earlier AC15, was unusual for the time, incorporating four EL84s running in Class A mode. This made it less efficient, hotter, and hence less reliable than the more common Class AB design. But it had *tone!*

The first examples had beige covering, but by 1963–64 the change had been made to the more common black.

By the mid sixties Vox had become a major force in the British music industry.

Shortly thereafter the company slipped into relative obscurity with a string of owners and second-rate products. But with the resurgence in things nostalgic, especially great amps, the Vox name was relaunched in 1993 with an authentic reissue of the legend, complete with chrome stand. In 1995 a limited run of purple AC30s was produced; it's a pleasure to show an example here.

The Vox Beatle, a.k.a. Super Beatle, is an impressive sight in any circumstances; but this 1966 NAMM Show special built out of clear plastic with chrome-plated internals almost defies description. This 240-watt model was one of the solid-state designs from the American branch of the company, not to be confused with the British Beatle stack, which was a 100-watt tube design. Complete with foot switch, stand, and covers, the black production-model amps retailed for a hefty $1,225 in 1966.

WATKINS

Watkins Electric Music, or WEM, was another British company that has since disappeared. Perhaps better known for their high-powered PA systems of the late sixties, they also built quite a variety of guitar combos, some of which looked pretty good. My favorite is the original Watkins Dominator, with a "V" front so prominent that it should have been called "Cyrano." Covered in two-tone blue and white vinyl, it was an 18-watt all-tube design with two speakers, and it really does sound good. Unfortunately, by 1976 it had lost most of the qualities that had originally distinguished it (see the advert).

The group shot includes, from left, a Scot, two Clubmans, a Dominator, and a Westminster. In front is the very desirable tube Copicat echo unit.

WAVESTREAM

The Wavestream V8 is without doubt the "big block" of amplifiers. With 22 tubes (including eight KT90s), they unsurprisingly weigh 100 pounds each—and you need *two* of them for stereo.

The brainchild of Scott Frankland, the V8 is a lot more than just a fantastic example of industrial design. The pyramid shape locates the heavy transformers at the bottom, while allowing convection cooling of the tubes through the middle of the chassis. The built-in meter allows individual biasing of each tube, as well as a choice of 150 watts of Class A or 300 watts of Class AB output. A patented "supercharger" circuit increases output by 8 watts per tube.

At $24,000 a pair, the V8 won't be within everyone's reach, but for those technically inclined, Wavestream publishes a white paper that describes the theory behind the design and makes quite fascinating reading.

WAVESTREAM

WHITE

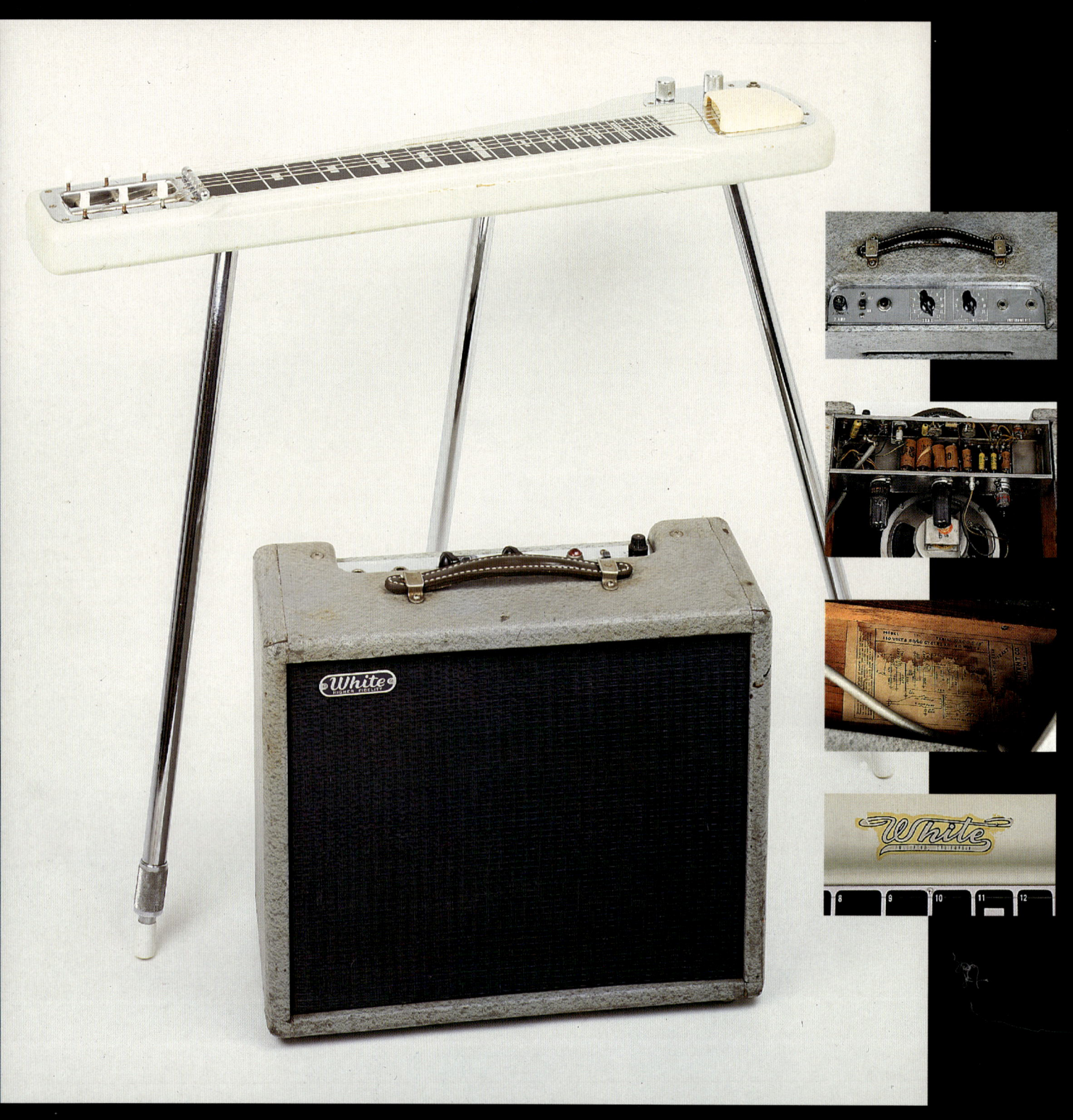

Forrest White, who died in 1994, played a major role in the history and development of the music industry, having worked for such companies as CMI, Rickenbacker, and Music Man. But it was as general manager of Fender from 1954 to 1967 that he is best known. His contributions were significant and did not go unnoticed by Leo Fender, who surprised Forrest in 1955 by building a new line of student steel guitars and amps under the "White" brand. Basically the same as a 1x8 Fender Princeton, the amp was a three-tube design, including a 6V6GT in the output stage.

The logo itself is a verbal treat: not content with high fidelity, the White boasts "Higher Fidelity"! How appropriate!